HVAC Controls 1.1

Norman A. Kellyman

ISBN: 979-8-9868821-9-2

DEDICATION

To all the people who work in the field of HVAC and HVAC Controls

Image Below (Standoing In Front of the Chiller System)

CONTENTS

HVAC CONTROLS SPECIFIC STYSTEMS TRAINING

ACKNOWLEDGMENTS

I would like to thank all those who help me along the way in the HVAC
and HVAC Controls field.

Image Below (Chiller System)

BOOK CHAPTERS

This book is divided into 3 parts.

Part 1 - HVAC

I believe a good understanding of HVAC is important before trying to understand Controls or HVAC Controls.

 Part 2 - HVAC Controls Basics

Understanding basic Controls principles can help   Once we understand Control principles then controls become much more friendly and easier to work with or trouble shoot.

Part 3 - HVAC Control Systems

This part of the book is impossible to teach since there are hundreds of controls systems. However we will provide some control training apps for Niagara N4, a very popular and robust controls system being used heavily on the market

HVAC 1.1

FREON, 4 MAJOR COMPONENTS DEFINED– BRIEF OVERVIEW

We have talked about the Removing heat from the desired location (be it living room, bed room or office space) by means of The Evaporator and getting rid of that heat by means of the Condenser generally located outside.

What we have not discussed how do we get the heat from you living room into the Evaporator and then out the Condenser to be dissipated. How do we continue this process until the space/ area needing to be cooled reach it's desired temperature.

Image Above (Freon Recovery Tank)

Freon

Well to remove heat from one location and transfer it to another require the use of a liquid called Freon or refrigerant. You've probably heard of the term in reference to global warming. A few years ago Freon was so inexpensive so dumping it in the atmosphere while repairing a defective AC unit was considered no big deal – until we began learning how it has contributed to global warming. Not to get you off the subject being discussed but know that today dumping any kind of Freon or refrigerant in the atmosphere is illegal. How the process works is, Freon absorb heat from the desired location while flowing through the evaporator than flows to the condenser and gets rid of this. To do this the Freon acts like any other liquid – it changes chemistry to absorb and then reverse it to dissipate heat.   Freon is a big item when trying to understand Air-conditioning and how the process works, for this reason we will get back to it later.

4 Major components defined

But for now lets talk about the what are the 4 major component that is needed to provide Air-condition or any type of refrigeration. We have already identified two (2), the Evaporator and the Condenser. The other two (2) are the Compressor and the metering device. These 4 devices are absolutely essential for any Air-conditioning or Refrigeration to take place.

There purpose and how the accomplish their goal in keeping us cool is next subject for discussion.

The 4 major component of the refrigeration/Air-condition cycle as was mention are the Evaporator, Condenser, Compressor, Metering Device. These work in conjunction with the Freon/ Refrigerant that was mentioned earlier to proved cooling our home, office or ever automobile. The Freon/Refrigerant absorb the heat with the help of the evaporator and release this heat with the help of the Condenser, this process continues until the set temperature has been reach – normally known as the set point. To do this work of going back and forth the Freon enlist the help of the two (2) other devices namely the Compressor and the Metering device.

To remove heat the Freon/ Refrigerant need to be 1. moving between the Evaporator and the Condenser to accomplish it's purpose of heat transfer. 2. Change it's state or consistency and then reverse back to it's original state to continue the process. *How does these four (4) major components come into play?* The Evaporator and the condenser are responsible for helping the Freon change state/consistence to accomplish it's task of

absorbing and removing heat. The Metering device helps the Freon get back to it's low temperature state so as to work at it's maximum potential of heat distribution. The .Compressor is what creates and maintain movement, it makes the Freon mover from each component to the next like the heart in human body – without which cooling will stop.

Knowing the four (4) major component of the Air-conditioning/ Refrigeration system and having some understanding of how the work leads us the to our next questions. *How does the Freon flow through these component?, when and where does it change stat?, why is change of state necessary for cooling or heat transfer?*

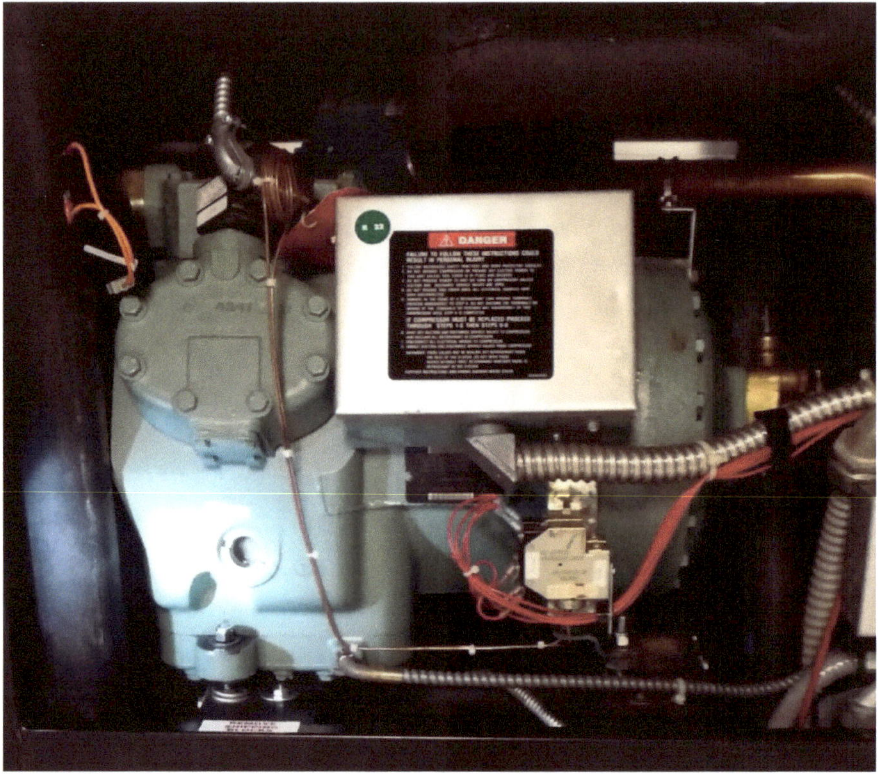

Image Above (Compressor)

COMPRESSOR

The compressor is responsible for keeping the Freon moving. Thus creating refrigeration of cooling. How does it work. It pulls in the Freon and pushes it out at the other end while at the same time increasing it's pressure. The compressor compresses the low pressure low temperature gas from the evaporator and releases it as hight pressure high temperature gas to the condenser. The makes the Freon temperature higher than the ambient (outdoor) temperature outside thus it will be very easy to cool 150 to 200 degree temperature Freon with 85 to 95 degree outside air.

Types of compressor

Compressor use various means to increase Freon pressure. Some use piston, like a car and presses the gas until it's pressure is increased multifold then releases it ot the condenser wile repeating the process over and over again. Other compressors use centrifugal force to compress the Freon gas then release it to the condenser.

Centrifugal force

Centrifugal force is what you feel when a car, train, or buss makes a turn. The force that pushes you the opposite side of the turn is centrifugal force. Multiply this force 100 times and you can crush almost anything. The compressor uses this force to compress the Freon from a low pressure low temperature gas to a high pressure high temperature gas.

Pressure

Pressure and temperature is directly related. When pressure is increased the temperature will increased as well. For this reason the compressor merely need to increase the pressure for the temperature of the Freon to increase.

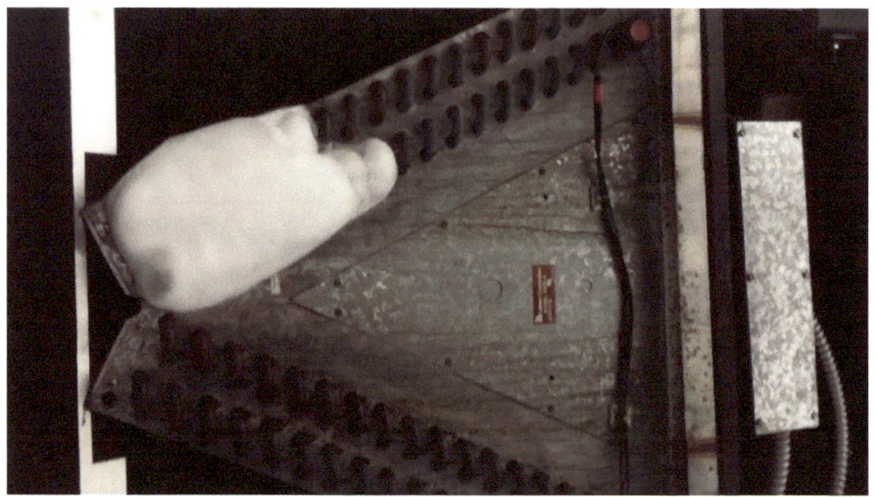

Image Above (AC Evaporator Coil)

EVAPORATOR

The job of the Air-condition/refrigeration Evaporator is to absorb/remove heat form the designated space. Wile removing heat it will reduce the humidity in the space. How does it accomplish this?

Cooling

By being colder than it's surrounding area the Evaporator absorb the heat from the space resulting in a more comfortable space as the heat is reduced. An understanding of heat and cold can help us grasp the job of the Evaporator for example – Cold is the absence of heat, the less heat the more cold you feel. The Air-condition does not provide cool, it removes heat. As a result we feel cool or comfortable as excess heat is removed.

Heat Removal

Air-Condition / Refrigeration is about removing heat. So cold is simply less heat, the more heat removed the colder we feel.

Evaporation theory

How then do we remove the heat?

Removing the heat In Air-condition or refrigeration involves providing a coil of copper or aluminum tubing and fins. This most of us have seen on the outside windows on the back of the widow Air-condition window unit. The Evaporator coil temperature is kept at a very low temp – about 45 degrees for Air-condition and 20 degrees for refrigeration. As the room air passes over this coil at about 75 to 90 degrees on a hot summer day. The evaporator coil absorb the heat form this are reducing it's temperature about 20 degrees. This process continues until enough heat Is remove from the room to reach what is set at the Air-condition unit.

Freon

Freon is a liquid with very interesting characteristics – unlike water which boils at 212 degrees Air – Condition Freon will boil at 45 degrees.   Why is this helpful? When a liquid boils and evaporates tremendous heat transfer takes place. The more heat we can remove at given time the better our cooling ability will be. For example – imagine yourself at the beach on a hot summer day temperature being about 85 to 95 degrees. After coming out of the water and standing in the wind you feel a sudden draft as the water begin to evaporate from your body. It's 90 degrees outside and you feel cold – this is power of heat removal as liquid changes state and evaporates.

Heat removal

The same process takes place in the Evaporator of the Air-condition unit – just multiply that chill several times over. With the Freon liquid capable at boiling at low temperature about 45 degrees for Air-condition and lower of

Refrigeration 20 degrees and below. Now as the room temperature passes over the evaporator at about 75 to 90 degrees, the Freon will boil and evaporate immediately absorbing as much heat as possible and dropping the this temperature to 55 to 70 degrees returning to space. Thus creating the chilling beach effect. Not only is heat removed from this are but humidity is removes as well. The process continues until the Air-conditioning unit is shut off or room temperature is satisfied.

HVAC 1.2

REFRIGERATION CYCLE TO REMOVE HEAT

How does the Freon flow threw 4 major components of the refrigeration cycle? What is the state or the Freon as it goes threw each component? How does changing state of the Freon effect cooling?

4 major components brief review:

As previously noted, to provide refrigeration or Air-condition of any kind 4 major components is absolutely necessary. These are the Evaporator the Condenser, Compressor and Metering device. As the Freon/Refrigerant moves though these devices it picks up heat and get rid of heat and the processes continue until you home or office reach the desired temperature...

Refrigeration Cycle (Freon/Refrigeration Flow)

Knowing how the Freon/ Refrigerant flow threw these devices is essential to understanding the refrigeration cycle. Without this understanding it is virtually impossible to understand Refrigeration/ Air-Condition theory. This knowledge is as fundamental as the 4 major components and we will need to know it as good as we walk and talk – it must become apart of us if we are to move further.

Lets start at the Evaporator. After picking up heat at the evaporator the Freon goes to the compressor, it then gets pushed along and enters the Condenser where it removes the heat it picks up in you home. The Freon/ refrigerant then enters the Metering device which prepare the Freon for maximum heat absorbing ability. So once again from the Evaporator to the Compressor to the Condenser to the Metering device and the cycle continues.. The Cycle is what's important not where we start, no matter where you decide to start one must understand the cycle.

(ex start at Compressor, Evaporator, Condenser, Metering device)

The Evaporator is where the Freon/Refrigerant enters as a Low Temperature Low Pressure Liquid, after absorbing the heat from your home it evaporates, this leaving the Evaporator as a Low temperature Low Pressure Vapor –Thus a change of state in the Evaporator.

Refrigerant State – Compressor

The Freon/ Refrigerant now enters the Compressor as this Low temperature Low Pressure vapor and as it's name implies the 'Compressor' Compresses This Low temperature Low Pressure Gas into a High temperature High pressure Gas. And send's it on it's way.

Refrigerant State – Condenser

The Freon/ Refrigerant now enters the Condenser a High temperature High pressure Gas and as it's name implies the 'Condenser' now Condenses this high temperature high Temperature gas/vapor into high temperature high temperature Liquid. And sends it on its way. At this point the Freon/Refrigeration is almost ready to pick up some more heat except it still need to get rid of the increased temperature to do so. This is where the metering device gives a helping hand.

Refrigerant State – Metering device

The metering device is responsible for decreasing the temperature and pressure of the Freon/ Refrigerant going into the Evaporator. Thus providing the Evaporator with Low pressure Low temperature liquid in order to absorb as heat again and continue it this cycle.

Refrigerant state in Evaporator and Compressor:

From the Evaporator as a low temperature low pressure liquid absorb heat vaporize and then to the Compressor as a low pressure low temperature vapor it becomes a high temperature high pressure vapor. Enters the condenser, condenses and leaves as a high temperature high pressure liquid to enter the Metering device which drops it's temperature and pressure to make it a low temperature low pressure liquid to reenter the Evaporator again..

HVAC – HEATING VENTILATION AND AIR-CONDITIONING ENTITIES

HVAC (Heating ventilations and Air-conditioning) is an extremely broad field. It comprises Ductwork, Air-condition, Cooling tower, Chillers, exhaust fans and controls to mention some divisions.   At times these different entities work together and sometimes not. Understanding the theory behind each division how and when they combine to provide cooling will help one grasp the sense of HVAC (Heating Ventilation and Air-conditioning.

A typical air condition can be small enough to fit in a window of ones house or larger then ones house, depending on the cooling need. After removing the heat form the space the Air-condition unit uses two mediums to dissipate that heat, Air or water. If air is used to remove heat from the Air-conditioner then it is a not combine with any of the above mention entities. When water is used to remove heat from the Air-conditioning unit it usually requires assistant to recirculat that water and keep it cool enough to continuer heat removal.

The cooling Tower is the Device Used to cool water use to remove heat from an air conditioning equipment. The cooling tower is designed to keep the water at a certain temperature, typically 80 degrees or so. It uses Evaporative type cooling, similar to what happens at the beach when we are cooled by the wind after coming out of the water, one feels a chill on a hot summer day because of the evaporated water. In the same way the cooling tower sprinkles the returning hot water from the Air-conditioning into the air only to have some evaporates cooling the rest and then circulating this cool water to be used by the air condition equipment once again.

As the building size and scope begin to multiple, for example a house to a hospital or 50 story office building. Chillers are the Air- condition devices used to improve Efficiency and cost in larger facilities. Chillers are very large capacity Air-Condition equipment used to chill water, which is then sent threw out the facility as the cooling medium. Cost is lowered because water cost less then Freon, and efficiency is increased as limitations to where we can cool decreases. Of course the chiller depends on the cooling tower to remove the heat it has picked up threw out the facility.

Ventilation is needed in critical and non critical areas   The Air-condition condition the air within the building, the better ones will also take in some fresh air from the outside and help remove some of the stale air that is being circulated. Some critical areas like Isolation rooms and Operating Rooms in a hospital need a steady streams of fresh outside air and to exhaust all air that is removed from the space. Professional kitchens need heavy duty ventilation along with some factories wile maintaining temperature.

Controls is either a blessing for the HVAC industry or a curse for the unprogressive technician. A large building can have anywhere from 10 to 1000 Air-conditioning units and more including chillers, towers, and exhaust fans. Controls brings all those units within reach of 1 pc computer. Reading temperature, commanding units on or off, increasing and decreasing temperature and if you cannot make it into the facility then all the HVAC equipment could be monitored over the Internet.

HVAC (Heating Ventilation and Air-condition) is a broad industry consisting of many different entities that at times combine to provide heating or cooling.

Image Above (DDC Panel with N4 Jace)

HVAC CONTROLS

To be a control technician one will need to understand HVAC sequence. Not necessarily HVAC theory. A good many control technician only know when to turn something on and off based on sequence given – generally called a sequence of operation. Knowing HVAC theory makes a control technician only stronger.

HVAC controls is about controlling and/or modifying Airflow. One need to understand and read what is called a flow diagram. This will change per job depending on the sophistication of the job. An airflow diagram list how air will flow throughout the HVAC system being controlled. Generally a flow diagram depicts the are coming into the Air-condition unit passing, dampers (mixed, outside) filters, coils (heating, cooling, humidifier), fans (Return and supply). This diagram is should fit the is usually created based on the unit given and the Sequence of Operation requested. It is also very helpful for troubleshooting.

Image Above (Johnson AX Panel)

HVAC CONTROLS OVERVIEW

The Goal of this Training material is to help technicians find their way around and troubleshooting most control problems. To that end we will talk about some commonly used terms in controls that may not be as familiar with an HVAC technicians. Understanding different types of HVAC systems and the impact such systems have on the control industry is vital to our understanding of today's HVAC controls demands.

Small appliances and larger air-condition systems have many things in common. People tend to shy away from thing as they become larger and more elaborate with more parts and wires. However I like to think of controls as a thermostat, as the system gets larger the thermostat expands. However in the end we will always do the same things like monitor temperature and turn the unit on and off.

Small to medium type systems and controls

Most people are not afraid of small residential system for homes of office buildings. An air-condition unit with mechanical cooling and a hot air furnace for heating is quite simple and can be understood with little time. However such system has within then an ample amount of controls, from high and slow compressor safeties and timers to Furnas or gas burner controls and finally our thermostat to controls the temperature of the unit.

Now as the home and office increase in size and there are more rooms it becomes increasing important to decide which route will be best financially to heat and cool that space. For a home window unit will suffice with a remote thermostat control while larger home with multiple rooms may opt for a split air-condition unit doing cooling or both heating and cooling. Zone dampers of multiply evaporator systems with one condenser will generally work fine for these application in which case the understanding of the zone system or multi evaporator system controls will do.

Larger system and controls

As the buildings increase size and the practicality of using window unites decrease, we need to consider other ways of cooling and heating wile at the same time properly managing such system and our resources. It may not be practical to have a building with 30,000 offices and 40 stories hight to have 30,000 window unit or even a large Freon type system sending Freon to the entire building. You can imagine the cost having an air-condition system with enough Freon to sent to an entire building or 30,000 offices.

HVAC CONTROLS OVERVIEW

CHILLERS, COOLING TOWERS, BOILERS AND AIR-CONDITIONS UNITS (DIAGRAM:)

To aid in cooling larger buildings and save on cost system had to be created to for this purpose. For these system to be economical, the medium of cooling had to be looked at.

In addition to the medium used to cool, the sheer magnitude of these system brought staggering cost of monitoring and maintain. Thus brining to light new names, and new terminologies like building management (control system).

Similar to the thermostat used to automatically maintain heat and cool while saving energy, these large system now required the same with as little man power as possible. Thus controls now became in effect building management system, designed to monitor unit and provide information to stationary engineers and other personnel. Such system can send alarm for critical areas and make phone calls when temperature is not what it is set to be.

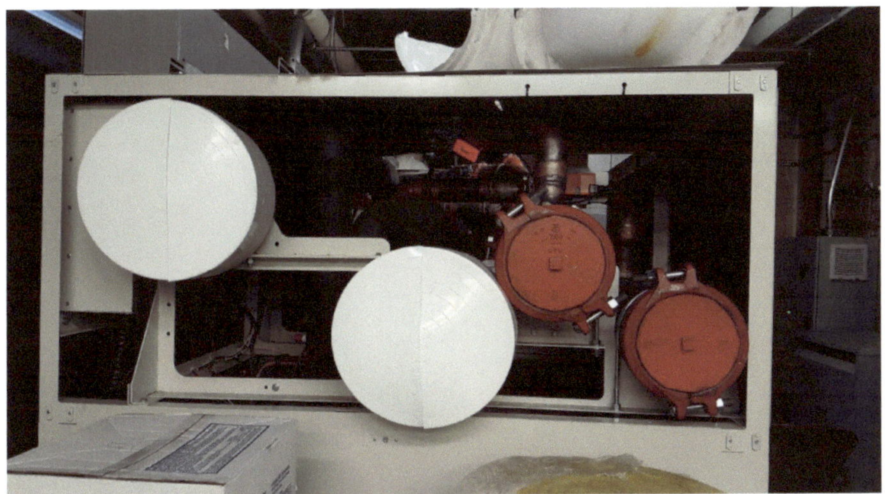

Image Above (Chiller & Condenser Barrels)

Chiller

Some of the system used to cool larger building are chillers. Basically a large air-condition system that use water to cool the building rather than Freon. The chiller cools the water then send the water to the entire building to use as the cooling medium. Water is cheaper than Freon and safer. When there is a leak, water will not injure humans nor ruin the earth like Freon and the cold of millions of gallon of water is a fraction what it would cost to send Freon through-out a high rise. Now that we are cooling millions of gallon of water to send through-out these high rise we needed a way to remove as the heat all this water is returning to the chiller. Here too the basic principle of air-condition did not change, only the system. Whereas in a window unit air-condition the heat is removed from the home and blow outside the air-condition, a chiller removing thousands of BTU at a time will need such a large condenser along with such good operating conditions it is not always practical.

<u>*Image Above (Cooling Tower Entrance Door)*</u>

Cooling Tower

To aid in removing such volume of heat the Cooling Tower now became practical. Using water with water cool condenser, the cooling tower can consistently send cool water to the condenser of the chiller, then blow air through that water and cool the water in an evaporation process as when are at the pool on a hot day and the wind blows across our body as we get a chill as the water evaporates. By using this process the cooling tower can consistently cool returning water to the chiller.

The demand for such level of cooling also meant a greater demand for such level of heat to units waiting to receive. Large boilers making steam or utilizing heat exchangers (making hot water from steam) would provide millions of gallon of hot water for these buildings .

Image Above (Low Pressure Steam Boiler)

Boilers

Similar to boiler in home that use oil or gas to create a fire and heat water or make steam for heating the home. Boiler 100 times the size are used to make steam in most building or heat water. The steam is then sent to an heat exchanger, a cross pass device without either medium touching but where the heat from one medium is use to heat the other, in this case the heat from the steam is used (exchanged) to heat the water. This hot water now is send through the building to be used as a heating source for the air-conditioning units. Some building used steam as a heating source, in which case steam is sent to the ac-units through-out the building.

HVAC CONTROLS OVERVIEW

VAV'S , CAV, VFDS

Image Above (Large AC Unit Using Chilled Water)

Large Air-conditioning systems

Larger space now need air-condition system with the ability to handle more square footage wile maintaining stable temperature. Think for minute about the air-condition window unite ability to cool the living room or the split air-condition system ability to cool the office. Now take that split unit or a packager unite and try to cool 300 rooms in a large office and now we have a problem with air-flow capacity and temperature issue. The last room will get no air, while the 1st room freeze with the feel of a tornado and so on.

Vav's, Cav's, VFDs

Larger air-condition systems need additional components to deliver the heat or cool temperature properly to their desired location in the space. Some of the components used in connection with Larger systems are variable air volume controllers (vav's), Constant air volume boxes (cav's) and variable frequency drives, fans with the ability to increase and decease speed.

Variable Air Volume (vav's), Constant Air Volume(cav')

Lets take the example above of a 300 office building and our need to provide cooling for all 300. Office number one needs - (125 *cfm - heat and 115 cfm for cool), office two need - (150 cfm - heatt and 125 cfm - cool) . We can now use variable air volume units, these unite consist of 2 things. A variable air volume box designed for this specific air flow and a controller programmed to maintain that air flow. Now we can maintain proper air flow for office 1 and 2 and so one. These unit can be used to control specific room temperate by controlling the amount of air to the space or by installing additional heat before the boxes to add or remove heat from the air being provided. When the box air flow is constant for heat and cool (ex. Room 3 - 130 cfm for heat and 130 cfm for cool) then the box is provided constant air in heat or cool mode and thus is a constant air volume unit.

Image Above (ABB VFD's)

Variable frequency drive

Now imagine running a marathon and then having to breath through a straw. Your lungs would simply bust at this critical moment because to the air pressure that has developed . There is too much air needing to be removed and this moment and the straw although good for volume of air will not work at this volume.

The situation is the same for the large unit with 300 variable volume boxes. In the mid day when the sun is hot and beating through the windows, all 300 variable air boxes are in cool modes and open fully. But as evening falls and things cool down you may have 250 variable air volume (vavs) units in heat mode and do not need as much air flow. To try to push the same amount of air flow as when all the boxes were fully open and calling for cool would cause an explosion in the air-condition duct similar to the marathon runner trying to breath through a straw at the end of a marathon.

Using variable frequency drives or fans that speed up and slow down as the pressure in the duct changes, we are able to maintain just the right amount of air needed at all times.

HVAC CONTROLS OVERVIEW

BMS, CONTROLS

Building Management Control Systems (BMS)

With Such large system serving entire buildings, it is understandable the our thermostat need to be expanded. Whereas a typical air-condition spit unit has 2 compressor for cooling, possible 1 heat coil. A chiller may have 12 compressor or larger low pressure chiller with the capacity of 100 compressor. We now need a system to control our chiller, our cooling tower(s), boiler(s), AC units , Vav's, Cav's, Vfds as well. This system should be able to maintain proper cooling tower water temperature for our chillers to run properly, inform us when our chiller is not provided the correct cooling water for the building before building begins to notice, while maintaining our hot water temperature and boiler control. We will need to know the temperature each air-conditioning (ac) units is supplying to the vav's or cav's boxes and wile monitoring the pressure of our variable frequency drives so as to have air available for all offices.

As the system become larger the parts needing to provide cooling are separated and Building Management controls are used to keep them together as one piece to save energy and maintain control. With a small window unite, all that is needed is a thermostat because all the pieces are in the window. The chiller on the other hand needs a means to remove heat like the cooling tower and this tower could be 100 stories up on the roof, however it is vital that the cooling tower water temperature is maintained and know by the operator as well as the chiller cool water temperature even though the

distance is so great. So too all the air-conditioning unites, variable air volume (vav's) and constant air volume boxes.

Controls

Using temperature sensor to monitor outside air temperature, tower water temperature and chiller supply and return water temperature. The building can be properly maintained and managed by a Building Management system that supplies this needed information to the operator constantly. All air-condition units can be monitored on the Building Management Control system similar to how computers are networked together and detailed information from each can be accessed. The level of control for each unit can be from simply monitoring (see if unit is running) to full control of unit (on/off, change temperature set-point, open/close status etc)

SOME CONTROL QUESTIONS

Why and When do we need controls?

Remote Access

To have remote access of HVAC systems and components. The ability to monitor and control from home, office or another part of the world.

GUI (Graphic user interface)

Controls and a large scale with many devices generally have a computer with graphics to make things easier and centralize. This Graphic User Interface (GUI) is great because is helps to decrease man power (1 person can see 300 unites and make changes) and save energy as adjustments can be made from 1 central location.

Miniaturize a large area or large amount of Units (150 Units on one computer screen)

Imagine the large building or several large buildings and the time it will talk check all the units one by one. A centralize Control system can decrease such time on a daily basis tremendously. It is not that a physical view of each unit is not necessary, on the contrary it is vital to check all HVAC equipment periodically and provide service. However normal day to day adjustment does not require one to visit every single unit.

Type of jobs and size of space

Generally Centralize controls with a Building Management Computer system is a large investment as many units are involved. The jobs are typically larger buildings or floors with much equipments. Although some critical areas with specific needs or critical temperature areas like clean rooms do require some type of automate controls.

VAV, CAV, VFD's

Vavs are short for variable air volume while CAVs are short for Constant Volume and they generally work with a VFD system which is short for variable frequency drive. These 3 are usually seen on the same system as the vav need the air flow to change the vfd need to be available to provide more or less air.

Can any Air-Condition system be fitted with Controls?

Yes , and as was mentioned, a thermostat is a form of control. However as the system become more elaborate so does the need for more or a control system with the ability to network, monitor and control larger system. A typical thermostat has limits were programmable Direct Digital controls can expand to meet each needs.

Types of HVAC' Controls Available on the Market?

A type of controls system and companies out there are

Honeywell, Andovar, Johnsn, Delta, KMC, Distect, Tridium.

What can be controlled on an Air-Conditioning System?

The things we control on an air-conditioning system are

Temperatures

We monitor the temperature and open and close valves or dampers, turn compressor on or off in order to maintain desired temperature. We can monitor and respond to temperature inside the space, outside or inside the unit.

Fans

Fans are turned on and off, speed increased or decrease to meet the need.

Imag e Above (AC Dampers)

Dampers (OAD, MAD, RAD {Fire dampers as well})

Dampers can be controlled to change air flow direction, minimize of allow more air to an area. Dampers can also be used for safety in the event of fire to stop dangerous smoke from spreading. Some or the reference to dampers are outdoor air dampers (OAD), return air damper (RAD), mixed air dampers (MAD), spill air damper (SAD) or fire dampers (FSD). These dampers all have different functions in a larger unit.

The return air damper (RAD) is control the air that returns from the space, we can take all the air from the space or limit the amount of air we return from the space to be re-circulated. The amount of air we do not wish to be returned to the space is then sent outside using the spill air damper (SAD) and we will also need to take in fresh air at the same time by opening the outdoor air damper proportionately. Fire dampers (FSD) close when there is fire of a fire alarm.

Valves or Actuators (cooling, heating valves) and Compressors

Cooling or Heating valves controlled by electronic or pneumatic actuators are opened, closed or modulated to maintain temperature control by the controls system. So too are compressors turned on and off as desired temperature is reached or needed.

HVAC CONTROLS INPUTS OUPTUS & BASIC LOGIC

INPUTS

What are Inputs?

An input device is used to bring information in from surrounding areas, things we need information from such as temperature, humidity, pressure, flow, current. On most Air-condition units we have similar inputs such as Outdoor temperature, return-air temperature, preheat temperature, discharge air temperature and space temperature. There are also Digital inputs or Binary inputs designed to monitor two position device as contacts, pressure switches and relays etc (Temperature, Fans Status, Damper Status, Valve Position, Compressor status, Static Pressure in Duct, etc.)

<u>Types of Input;</u>
Analog Inputs Digital Input Variables

Analog Inputs:
 Input sensors that can read value

    RAT - Return Air Temperature, PHT - Pre-Heat Temperature
    CCT - Cooling Coil Temperature, SAT - Supply Air Temperature
    OAT - Outdoor Air Temperature, RM Temp - Room Temperature
    Duct Air Pressure - Transducers Static pressure

Digital Input:
On/Off Input sensors or Input sensors that see a state

    CT - Current Switch
    SAF Sappy Air Fan, Return Air Fan, Variable Fan Drive State,
    AC - Unit on/ off

Variables:

Make believe inputs that hold values; variables could be both Analog and Digital inputs

Or

Set Points:

Dummy Digital or Analog Inputs to hold values and other Information

   Set Points
  OAT Changeover,
CoolStpt - Cool Set Point
  HtStPt - Heat Set Point
  On/Off change of state variable – no button but software state.
  Examples   LlStPt - Low Limit Set point

  Unit On/Off

HVAC CONTROLS INPUTS OUPTUS & BASIC LOGIC

OUTPUTS

What are Outputs?

An output will command a device in the field. There are Analog output devices such as drives, valve, damper actuators etc. Then there are Digital devices which are on / off devices such as current switch, fan status etc. (Fans, Valves, Dampers, etc)

Types of Outputs:

 Analog Outputs Digital Outputs Variables

Analog Outputs:

Output that send out values

 VFD-0-10 VDC signal to ramp up

 Valve Actuator – 4-20 milli-amp to open and close

 Damper Actuator: 0-10 VDC To open and close

Digital Outputs:

On/Off Output or Output that Change state (one/two 1/0 etc)

Fan start stop

Unit on/ off

Compressor On/off

Non modulating Dampers Open/Close (fire Dampers)

Non Modulation Safety Valves Open/Close

Variables:

Dummy Digital or Analog Outputs that hold Values

Software Timers On/Off modes other than hand switch

Stage Sequences – Multiple compressor system;

Compressors should com on at a time and when

Lead Lag – Signal next compressor or pump to go down

Note:

You will have to tell the controller what type of Input or output you want to use before programming. There are controllers with Universal Inputs and Outputs and all you need to do is just change a jumper and it becomes a different type. Some controllers Inputs or outputs are not Universal thus will not change.

HVAC CONTROLS INPUTS OUPTUS & BASIC LOGIC

LOGIC AND SOFTWARE POINTS

What is logic and Software Points?

The inputs and outputs will stay where they are until you or the software tell it to do something. This is where we need to program some type of logic sequence the allow these values to run smoothly and supply Cooling or heating to the space.

Automating the Output and inputs to provide Cooling or Heating, is like making the unit come a live or operating as if it was thinking.

Software Points are Mini programs that help speed up our programming process.

 To Aid in our programming there are various software points inside our controller available for our use in addition to the variables.

<u>Helpful Software Points:</u>

Loops
Stage Sequencer
Math Logic
(And, or, and/or, Greater than(>), Less than(<), equal to (=), etc)
Logic: using these Software points with some logic can give and Unit a brain.

<u>Basic Software logic example</u>

If a situation occurs you want a response.

Or

If a situation does not a occur you want a response

Example 2:

Room Temp Gets to 70 Put on Cooling.

LOGIC

If the Room temp is => (equal to or les than) 70 degrees Turn Output 1 to On

DIAGRAM AND PLANS

FLOOR PLANS

Floor plans are designed to give direction of the space like a map gives direction for city so does floor plans map our a building or floor of a building.

Location of air condition unit fan rooms, emergency stairs, ceiling access, duct work run with dampers locations and temp sensors location are all available on floor plans.

<u>Floor Plan Practice Example</u>
From the Floor Plan below locate AHU-5 , AHU-1 and ACU-5.   (Hint *they will be circle in red.*)

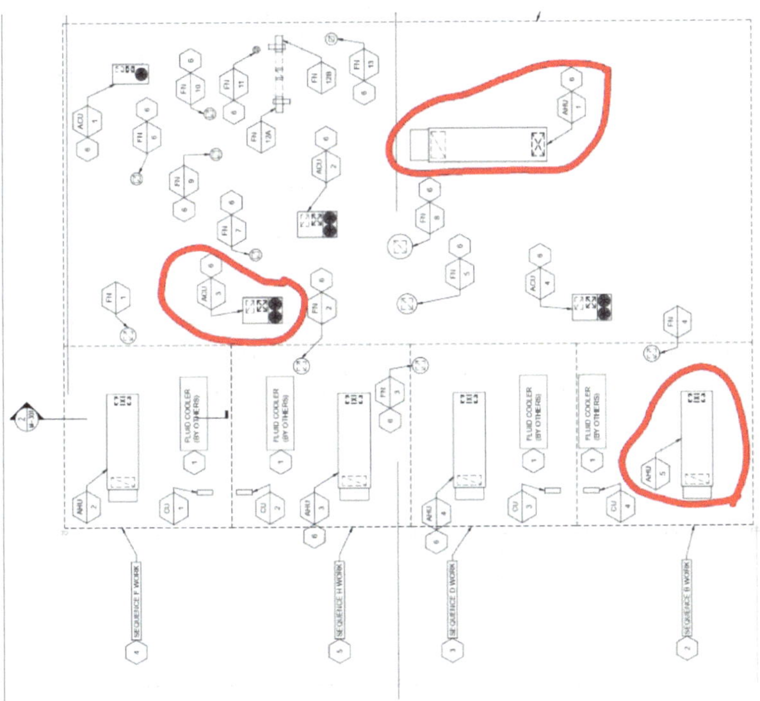

Image Above (Floor Plan Example)

WIRING DIAGRAM

Wiring diagram shows how all the wires are connected to make the controls work or the system functions. Troubleshooting system problem without a wiring diagram can be very difficult because some systems have hundreds of wire connections.

A strong understanding of wire diagram is important for anyone doing controls because most the HVAC Controls related problems involves tracing controls wires to find defective components. One who cannot trace controls wires will not be a successful control Technician. The ability to find defective components as a control technicians require isolating working parts until the defective part is found a skill that cannot be done without knowing how to read a wiring diagram.

The stronger your understanding of wire diagram the better control technician you will be.

<u>Panel Wire Diagram Practice Example</u>
From the Panel Wire Diagram below locate TS-1 HWS Temp, V-2 Steam Valve and HW Pump P-5 VFD Enable. (Hint *they will be circle in red.*)

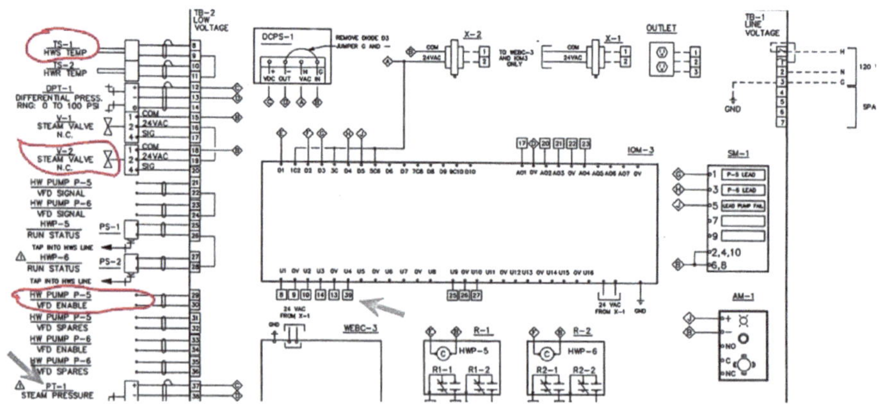

Image Above (Panel Wire Diagram Example)

FLOW DIAGRAM

Flow diagram shows how air flow threw the air condition system along with location of devices as the air passes. When working on a unit it is important to know and understand flow diagram in order to locate sensors, fans, dampers etc.

If you cannot identify how the air the air flow threw the air condition unite it is impossible to service the system because you will not know where to start.

Flow Diagram Practice Example
From the Flow Diagram Below locate RTU-1 Air Flow to 3 VAV's Downstream. (Hint *they will be circle in red.*)

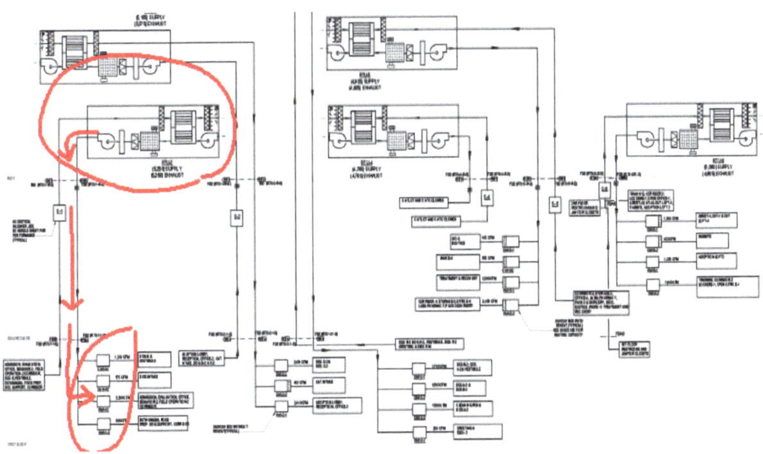

Image Above (Flow Diagram Example)

HVAC CONTROLS TRAINING

https://nakenterprisecontrols.com

<u>HONEYWELL N4 TRAINING</u>

ABOUT THE AUTHOR

II have been doing HVAC and HVAC controls now for over 20 years.